ROGER THARPE

JOURNEY THROUGH THE DESERT

One Soldier's Journal during the Afghan War

vidi press

Copyright

Journey through the Desert:
One Soldier's Journey during the Afghanistan War

ISBN: 978-1-934561-01-0

PUBLISHED BY

Vidi Press
11721 Whittier Blvd. #203
Whittier, CA 90601
vidipress.com
800-409-7170

COVER DESIGN AND INTERIOR LAYOUT BY

LionsgateBookDesign.com

Dedication

★ ★ ★

This work is dedicated first and foremost to the men and women of the battlefield—those families who have lost loved ones on the sand and to the many friends of those wounded warriors. This work is dedicated to the spouses who have military members and civilians in the desert fighting for our freedom and the freedom of those who are unable to free themselves.

Preface

★ ★ ★

The Lord is my shepherd; I shall not want.

~ Psalm 23:1

The above psalm was written by the great King David, a man who won his fair share of battles and also lost some battles. King David was a warrior, empowered by the God of the universe. His relationship with God, or God's relationship with him, gave him the victories. David writes that the Lord is his overseer and his guide, bringing him to the point of not wanting.

This journey through the desert from September 2008 until April 2009 was written under the umbrella of not having a lasting want, for the Lord was my shepherd. As the fighting went on and funerals went on and prayers of encouragement and blessing by chaplains went on, we continued to look forward to another day. We never knew when our day would come to an end or our life would otherwise be changed forever, but my hope and others' hope came from our prayers and from church members who sent care packages and from school children who wrote letters.

These gifts would aid in our survival and in our joy. Through it all, I found Christ to be the linchpin in my success and return. It is not as though my return reveals Christ's absence in another, it only reveals that through all things whether in the desert or in safety that the Lord is my shepherd and I shall not want.

Contents

★ ★ ★

Dedication.............................iii
Preface.....................................v
Introduction.............................1

★ September.......................... 5
★ October 9
★ November 15
★ December..........................20
★ January.............................34
★ February 48
★ March.................................62
★ April 75

This Journey Ends..........79

Introduction

★ ★ ★

From the womb, I struggled not to be removed. The doctors told my mother to push; and as she pushed, I held on. Desiring not to leave the safety of the womb and its familiarity, I held onto whatever I could grab. Despite my best efforts, the unmerciful prongs of life got a grip on me and the will of God overcame my objections. New life entered the world—a world that I didn't ask for; and yet, I was here.

The amazing thing about life is that the journey has already been taken, the concrete road has been laid, and the path has been declared. Some wonder in glorious awe at this life and others hate the very breath they breathe. But why? We all have been created in God's image, according to the first book in the Bible. The design is his. The timing is his. The opportunities presented to us are his. We do not choose anything, of what I would consider to be, of any great proportion. We are designed and planned by God and move according to his plan. "Who can make straight what he has made crooked?" (Ecclesiastes 7:13).

I ponder the magnificence of a God who has not only desired my creation but also manufactured my parts and programmed its circuits. To discover His plan, one component at a time, illuminates my soul. Once it was all hidden, but God continuously reveals himself through the ever-growing relationship I have developed with him.

I recall a Scripture that says his secret is with the righteous. God reveals more and more to those who seek him. That is why the Bible says that those who seek him will find him. But it doesn't stop there. God rewards those who diligently seek him. Moses asked to see God; and so, God revealed his backside to Moses.

If we would take action on every precept that flows from Christ, he would reveal more and more truth to us because his secret is with the righteous. Why do you think many in Scripture knew when they would die? Because his secret is with the righteous. Not saying you want to know that information, but if God is willing to reveal the end, what else is he willing to reveal to those who diligently seek him?

The home which God has placed me in is significant for the journey I must travel. The journey has been tedious and difficult but is not unlike yours; we all have a difficult story to tell. Each one, though, is lived so that the purpose of God will be fulfilled.

One thing I have come to know is that God chooses the base things of the world to confound the wise. I have spent many years being the object of my classmates' humor. They would laugh at the way I answered questions or the comments I made when a teacher or classmate did something odd. I recall when my earth science teacher showed us a picture of his wife in a one-piece bathing suit. He chuckled and said, "She was a beauty." I said (judging by his aged appearance), "She doesn't look like that now!" The class laughed and he quickly placed the picture back into his wallet. The students tried to get me to say other things about her so the laughing could proceed.

My classmates thought of me asa jester and my mother hated it. She worked hard to raise a child who would be taken seriously. But I wasn't. Often, I could hear my thoughts reverberating the message of the kids talking about me. "There goes Roger again," they'd say. "Tryin' to walk inside the lockers." The school hall was a scary place for a boy with severe social anxiety and attention deficits, so I walked as close to the walls as possible, avoiding my classmates as best I could.

Come to think of it, Basic Training wasn't much different recruits made fun of how oddly I marched. As you can imagine, I became the object of at least a few soldiers' jokes.

Throughout school, I failed miserably with my female counterparts as well. Once, I gave a rose to a girl on the bus and sat next to her. My friend urged me to speak to her but, of course, I stayed silent. My success in the classroom wasn't much better. In spite of my learning difficulties, I made it to senior year of high school, but would have failed if my sister didn't ask one of my teachers to give me a second chance on the final exam.

My attention deficit disorder impaired my ability to develop socially and academically but,in the midst of all of this, God spoke to me.

I first heard from the Lord when the fear of lighting and thunder caused me to flee the warmth of my bed and run to my mom's arms for safety. At this point, many parents might have told their kids "There's nothing to be scared of. Go back to bed and go to sleep." But not my mom.

"Have you spoken to the Lord about the storm?" she asked.
"Nope," I said.

"Well, get back to bed and speak to God."

So I got back in bed and talked to the Lord about my fear. Immediately, a thought entered my mind—one of God's thoughts from Psalm 56:3: "Whenever I am afraid, I will trust in you." It worked better than Advil.

Through all of the laughs and the constant desire to remove myself from the world, I am still here. I still struggle with difficulties of Attention Deficit Disorder (ADD) and the intense social anxiety that I have secretly and not-so-secretly endured throughout my school and military career. But with my ear close to his chest, I am able to hear his voice. Jesus has made himself known to me and I am ever grateful.

What follows is my journey through the desert of Afghanistan as recorded in my personal journal from September 2008 to April 2009, a spiritual journey of God's leading, loving, and teaching. I did not write it only for myself but as a legacy for my children, when they are old enough to understand it; for my wife, so she knows how much I love her; for my family and friends, so they know how much I appreciate their support; and for you, so that you will know the wonderful providences of God. I also hope that some pitfalls and traps on your own journey might be avoided through the listening and assimilating of mine.

I have since added some explanatory notes and other details to make this journey readable to readers who may not be familiar with military jargon or ways of life.

September

★ 28 Sep 2008

It seems Skype is acting fuzzy. I'm not sure whether it's the program or the poor connection or combination of both, but I think God has something to say about this.(For those who are not familiar with Skype, it's a computer program used to connect to others over the internet by voice and webcam.)

I was reading the gospel of Matthew earlier today and found that the prize goes to those who endure all things, even internet glitches. I don't know if theologians would agree; but keeping that in mind, all is well. God is good all the time. His mercy endures forever.

Changing gears a little. The weather here is hot and the smell is putrefying, like sulfur. It covers the whole land like a mask covers a face. The great winds don't let it settle either. They blow the stench right up my nostrils.

This environment makes me intimately aware how evil man can be. There are so many resources in the world, but because of the greed and pride of humankind it cannot be shared. God must come and restore what we have destroyed in our disobedience.

At the same time, I am reminded of the book of Proverbs 16:4 where it says that even the wicked are in God's plan. The Scriptures tell about many stories and many paths that lead to everlasting joy and peace and everlasting hell, torment, and damnation. Notice that I mentioned only two attributes concerning our eternity with God and three concerning the separation from him. This is because the amount of Scriptures regarding the punishments of God outweighs the amount regarding his rewards. God wants to warn us in every possible way so that we can escape the pending judgment. That is why he not only gave us his Word, but also preachers.

"You donot have because you do not ask," says James 4:2. I decided to tryout this procedural proverb today; I was told that I and other troops would have to carry my own bags on the flight to Kandahar. The terminal is a few blocks away and carrying 50-pound bags—which I have four of by the way—is crazy. So I prayed to the Lord for some kind of assistance and a funny thing happened. The moment I exited the porta-potty, a vehicle drove up in front of my hut.

Even though I was told not to ask anyone for assistance, I followed the inner promptings of peace and desperation and asked this man for assistance. He gladly offered to help. As we were driving, he shared with me a message on joy that he had just heard. He was a Christian and spoke encouraging words concerning the journey that lies ahead for me in the desert.

Jesus tells us to ask and it shall be given to thee. Our words exit from the mouth and, based upon the promise of God, he will answer with a word. For out of the mouth of God his Word creates the manifestation of our desire. Through his love, he sends his Word to accomplish that which he desires for our Father. All things work together for the good to those who love God.

I have to go. Will write some more later. By the way, do not hesitate to ask God for specific requests. We have not because we ask not. Let's ask and let God decide whether or not he will provide. My prayer is that God will shorten my tour to four months. Whether he moves or not, it will not be because I did not ask.

★ 29 Sep 2008

Kandahar is 8.5 hrs. ahead of Texas and 9.5 hrs. based on daylight savings time. With work hours from 1700 at night to 0100 in the morning, it's difficult to reach my wife on the phone. She's either at work, church, or the gym. Time drags when I cannot talk to her.

I've been sitting here, reflecting on my life and how much pain has filled it due to my learning abilities and even wetting the bed, which I didn't stop until I was a teen. Sometimes I think that God has really given me the short end. No one likes to be struggling when it seems that others are not. I thank God for my parents, though, who took the time to get educated about my problems and not be in denial.

There are times when I feel that my choosing to enter the military was a mistake and that if I had a greater processor, meaning faster mind, this choice would not have been on the menu. But as I read through the Old Testament, I see that God sets the scene, gives us his Word, and it's our responsibility to follow it, even call him on his Word. This sounds bold, but many examples of people doing this are in the New Testament.

I recall the woman who continually visited the judge to make her request known. Lest she weary him from her constant badgering, he granted her request. I must also keep coming to God, but I must be certain my desire is best for him.

I was meditating today on Psalm 37:4: "Delight yourself also in the Lord, and he shall give you the desires of your heart." I think sometimes as Christians, we use God as an excuse not to do things that he intended for us to do, like getting married or taking a particular job. But then I think of Ruth who took action and went after Boaz. And I think of those engaged in battle, like David taking charge to meet and fight Goliath. He wasn't just standing around praying about it. I need to go after my desires, too, but much of my problem is learning the line where I end and where God begins. My life is not my own and I have been bought with a price.

Hezekiah, the king who dared to destroy idol worship in God's land, didn't want to die. He pleaded with God to save his life from his life-threatening disease. God granted him a fifteen-year extension; and yet, as we look at the rest of his life, death would have been the better option for him. If he had died earlier, he would have left the earth with an honorable discharge similar to Enoch.

God knew that during those fifteen years, at times, it would have been best for Hezekiah to die, but because God could only follow his Word, He had to let Hezekiah live.

You and I often pray according to God's word, but are we believing that he has to do it?

God is constant.

God cannot lie.

October

★ 2 Oct 2008

A troop of mine has just come back from a mail run and offered me an opportunity to distribute food and clothes to Afghan children "outside the wire." That's military jargon for off–base. As I ponder the opportunity, my heart says, okay, but my mind says it's not a good idea. My personal desire is to avoid having to engage someone in battle. I mean, for real. Those convoys can be life-ending. Is this too honest a statement? Don't get me wrong. I signed up to do a job and will defend my country with my life, but doing so is not a desire. I do not believe it is any soldier's desire to die in battle; he does so in obedience to his commitment to support and defend.

My oath of office affirms my commitment to protect the nation. But according to the Scriptures, I am supposed to present my body as a living sacrifice, holy and acceptable to God, which is my reasonable service (Romans 12:1). It is in the expectations of these two requirements to serve where faith is tested against wisdom. Am I willing to present myself a target to the enemy in exchange for helping Afghan children?

At present, I am not.

★ 4 Oct 2008

No matter how much I kick up the dust and complain, the Lord has his way. In the whirlwind,it's clear that his way is not my way. I grow disillusioned by the thought that I am able to easily believe in the cross of Jesus Christ and yet a simple prayer for a shorter deployment or the war ending sooner is performed with great difficulty. Is this because I doubt that God can handle the size of such a prayer or that I think my absence as personnel will greatly impact flight plans and create political turmoil? During times like these, I wonder if I really believe the stories of God's intervention within the Old Testament, such as when King Jehosaphat prayed to the Lord about the impending destruction of his nation by neighboring kingdoms. Without his army lifting a finger to fight, the opposition fought among themselves and the king's land was saved.

Story after story written in the pages of the Bible reveal how the prayer of believers seemingly moved the hand of God, through his heart, for their deliverance. And yet, sometimes, I fail to believe.

Of course, there is danger in thinking God *must* answer my prayers in the way I pray it; he is God and ultimately does what he desires. So I have resolved to maintain my prayer for a specific answer to my request for a shorter deployment and be satisfied when I receive it. I know God hears me and whatever length it turns out to be, I am certain that will be his answer.

★ 5 Oct 2008

Being away from my wife is the cause of my grief; for if she were here, I wouldn't nearly feel the disgust of this deployment. I consider why I am here and she is there. What God makes crooked who can make straight? The temptation to always have an answer to this question pulls at the very depths of my heart and, yet, in the words of Solomon, it's vanity and vexation of the Spirit. What I truly long for resides in the fact that I am not yet complete. I was joined together with Christ by his blood; but since I have not yet been fully transformed to be like him, I sigh.

The book of Revelation says that even the elders in heaven continue to ask the Lord, "How long?" In the meantime, I realize that it's of the Lord's mercy that any of us haven't been destroyed and that I'm here at all; his compassion for us never ceases.

★ 7 Oct 2008

Many thoughts are running through my mind as I think about a foreign national I met today. He believes in Jesus, the Holy Spirit, salvation, and baptism, but has strange thoughts concerning the inerrancy of Scripture.

He says that Scripture contains the Word of God without being the Word of God. He says the Holy Spirit gives information apart from the word. I told him that I agree that the Holy Spirit sometimes gives wisdom for which there is no scriptural reference, like which home to purchase. However, if the Bible only *contains* God's word, that allows my physical spirit to choose what feels best rather than doing what Scripture says is best, even if it doesn't feel right.

He had quite a vocabulary, too . . . colorful curses that would make a sailor cringe, ignoring the Scripture verse in Philippians 1:27(KJV), which says to"only let your conversation be as it becometh the gospel of Christ."

He told me that when he was explaining the gospel to non-Christians that his jaw would turn to concrete so that a man who was angry at him for witnessing would break his fist on his face. I do not deny that God is able to and does perform miracles, especially for the purpose of witnessing, but I perceive his story is based solely upon a unique personal experience (even if it really did happen) rather than Scripture. This approach is problematic for a couple reasons: First,the devil has power to perform supernatural works, and the Lord Jesus tests us to see who we will follow.

Second, relying on personal experience alone produces a shifting foundation—from God's Word to our own words. Different experiences judged by different experiences may or may not be correctly interpreted, and so they can lead us down a path of feelings. These feelings may produce pleasure and a feeling of self-righteousness but, actually, they create a chasm between us and God. The solid foundation (the rock) we must stand upon is Jesus Christ. The Bible given to us is for our correction, reproof, and doctrine so that we may be full in the wisdom of God. If the foundation is removed how shall we stand?

More and more, I find myself thinking like Christ concerning people who desire to know the truth as well as those who are simply satisfied in presenting views of which they themselves have not studied. It's my hope that my encounters with this foreign national will produce in him a zeal for the real Christ, not merely a mental exercise arguing about topics that have little to do with Christians edifying each other.

Times like these, I think about how Jesus related to the religious leaders, the Pharisees. He would not give them an inch in their monologues but, instead, spoke to them sharply. He called them hypocrites and white-washed tombs (clean on the outside, but dead on the inside). He simply did not waste much breathe on them as he knew they would not believe him no matter what he said.

The Pharisees tried to stir up strife among the people and to make Jesus look as though he was not who he claimed to be. My encounters with this fellow confirm in me that God has not commanded me to argue or strive for the gospel either but just to tell the truth whether or not he is receiving it. So I do.

My encounters with this foreigner also confirm in me that whatever happens to me here, it is for the furtherance of the gospel of Jesus Christ. As mentioned earlier, we are not our own; however, the liberty that I have with Christ, who has set me free, enables me to have a personal relationship with him. That relationship allows me to call upon him with specific requests. We must not forget in the pursuit of living for Jesus that he has made available a friendship that says, "Ask and it shall be given."

When dealing with co-workers who would like to see me terminated, who plot and scheme for my removal, I also need wisdom. I need to run to God. I have done so and while others around me have been terminated and are no longer in the service, I remain. Not because of my intellect, remember. I have ADD. It's God.

This brings to mind how Jesus taught the disciples to grow mature in Christ. He constantly placed challenges before them. He sent physical storms to teach them about the true identity and power of God through Jesus. He even gave them authority to cast out demons and then allowed them to fail at it so that they would understand even more about God's ways.

The challenges that are brought my way cause me to flee in prayer to the Lord.

He often places individuals in my life, like this man with a wishy-washy spirit that cause me to run to him for knowledge and correct advancement. I was in the gym recently and there came in a soldier with an able body, but the area where a mouth was supposed to be was so deformed it was nonexistent. I heard the Holy Spirit:"Heal that man." I was on the leg lift machine working my abdominals and again: "Heal that man." By now, the voice was bothering me, so I stopped working out and ran to the bathroom. I asked the Lord, "Is this you? If it is, then make the water run in the faucet."

The water ran.

Like Gideon (who didn't believe God's answer either), I still wasn't convinced. I was also afraid that nothing would happen; that instead, I would be looked at as a fool and endure even more joking at my expense.

I didn't move at all until the man left.

I failed that test.

November

★ 25 Nov 2008

There is but one God, however many things called god. Some worship crystals, mystical practices, weapons, angels, popularity, wealth, and/or themselves. They wonder.

Mystery God where are you?
Are you in the clouds?
Are you in the stone?
Why the confusion for us to find you, Jesus?

Hmmm.

Did you notice the switch?

Notice how the questioning begins with God and then the name of Jesus pops up? Why is this? Could it be that the inner Spirit knows Jesus is God? If some people know Jesus is God, why don't they also believe that Jesus is the Savior?

The next time I get a chance, I'm going to explain to a worshipper of stone how Jesus existed before Adam. If Jesus existed before Adam, that means Jesus was God before the creation of the earth and he was God when he came to earth 2,000 years ago and he is God in heaven today. He proved it, too. Jesus forgave sins on earth, and only God can forgive sins. This is why there was so much anger from the Pharisees in Jesus' time and from religious leaders today. But when I get a chance,I'm going to explain to them that there is no name given among men but the name of Jesus. In the name of Jesus, we receive the restoration of all things because he loves us.

All things created lead back to a word. This word is love. Why love? The element that many search and seek after is love. This desire to be loved begins in the very early stages of life. It would appear the only things needed are food, shelter, and clothing. But studies show a child who isn't held will die. Why?

God reaches out with his hands through his disciples, ministers, and children to a world desiring to be loved. I love you, I love you; Love me, Love me.

You desire the truth. Jesus says, Call my name and believe; they that call upon the Lord shall be saved. Ask him today about this mystery. If doubt is present, know that doubt did not prevent Thomas, who needed to see Jesus' wounds, from seeking the truth.

It is in seeking the truth where the supernatural ear of every creature hears. Make sure you are serious. This is a serious request and a serious action that will garner a serious response. Call today and call often. Jesus is not a man that he should lie.

He promised that he will reveal himself to those who ask, and he has never changed that promise. He is from the beginning and will be present in the end.

★ 30 Nov 2008

I am writing today to recount a conversation I had with a believer, one who professes Jesus as Lord and Savior but does not put much value on the Word of God. He asked, as many do:

What is the purpose of life?

What is required to live life properly?

Is wisdom enough for a complete and successful life?

Does the end of life make sense?

Not in too many words, but fairly given,here are my answers to the questions, followed by his response.

I explained to him that *the purpose of life* is fulfilled in the Creator of life. From the beginning God created life and sustains life for his enjoyment. But God was grieved when the pure and perfect relationship he planned to experience with his creatures had been ruined with their disobedience.

The purpose of life is now seen through the canvas of spiritual corruption for which true visibility is found in Christ alone. I told him that this corruption would have separated us from God forever, but he sent his Son to save mankind by putting us back in right-standing with him. This allows us to escape hell, which was originally designed only for his fallen angels.

We live life properly through the wisdom attained by the study of the Bible and through the comforter (the Holy Spirit) from whom we receive revelation to apply the wisdom that we learn in life situations. As God guides us with his hand through time, experience, and submission, we learn more and more what it means to live the way God intends. We live in power as we stand in unison with the King's command. For where there is the Word of the King, there is power. This power is used to love our families, neighbors, and enemies so that ultimately souls may come to Jesus and live.

Wisdom is not enough for living this life successfully. Solomon was given supernatural wisdom by God and yet lived a promiscuous lifestyle. This man had several women for each day of the week; seven hundred wives and three hundred concubines—additional women Solomon would enjoy. These women turned his heart to the gods they served. How do you think this happened?

The Word of God tells us *to watch*, which in Scripture is *to pray*. Those who diligently spend time with God in prayer have power to apply the wisdom attained.

As believers, the possibility of reaching perfection, in the sense of being flawless, is impossible. But we may succeed at deterring the majority of mishaps that happen to even wise men and women of God, like Solomon, by spending hours in prayer every week.

The end of life makes perfect sense when you understand that God promised Adam death for sinning against him. And as it is appointed unto men once to die, but after thisthe judgment (Hebrews 9:27), the only way to enter heaven—the place of eternal rest, peace and tranquility—is to die in right standing with Christ.

It has not entered the heart of man what wonderful joys are in store for God's sons and daughters in Christ. Heaven makes the end of life make sense.

After all of this explaining, he insists that none of what I said matters and that the Holy Spirit is all that is required for life. As the tone of our conversation increases and the heat behind our differing opinions begin to rise. I go on to tell him why the Bible is so important for life.

I argued that the Bible makes the purpose, pleasure, and end of life make sense. I said that in the beginning, God is. God spoke a word and created life—through God's word, the world exists; and with God's word, he created man in his image. The image is the word. We cannot, nor has anyone ever, seen God; and yet, we are made in his image. The Word of God draws us near to his image—to him.

Words are so important. Words create buildings and destroy them. Words create peace and also create war. Words create books which change the minds of civilizations. With one word of a king, people die. With the word of a scorned person, people die. The spoken word is faster and more easily entreated than the written but, in many ways, the written word is more powerful. For instance, our laws are written words and ignorance of these words isn't excusable. Why? It's because once on paper, they are binding. If the spoken word in man were sufficient, there would be no need for books and no need for God to have written down his commands on tablets of stone or in laws, chronicles, poetry, letters, and gospels.

The Word is the most powerful thing on earth. "In the beginning was the word, and the word was with God . . .and the word became flesh and dwelt among us" (John 1:1, 14). This message will add to knowledge already held or forgotten. In some way, this written word once it enters the mind will move the reader to either act upon the word, change its meaning, or discount it as foolish blunder.

He replied to all of this by saying that since we are in the desert and at war where it's not always feasible to tote around a Bible, what does a soldier do without one? I answered with what King David said in Psalm 119:11 (KJV): "Thy word have I hid in mine heart, that I might not sin against thee." Needless to say, we both left unmoved from our previous position; however, in secret, I believe he was moved.

December

★ 4 Dec 2008

I was thinking how 1 John 4:8 says that God is love. And Paul mentions in 1 Corinthians that the greatest of all the gifts is love. Love doesn't require a degree in theology, secular humanism, or even pop psychology. Through faith in Jesus Christ, we only need to love him with all our heart, mind, soul, and body.

Our love for Christ manifests itself through our love for those with whom we disagree, for our enemies as well as our friends. While love is the greatest gift, wisdom follows close behind. A double-minded man is unstable in all his ways; but even for him, there is hope: "If any of you lacks wisdom, let him ask of God, who gives to all liberally and without reproach, and it will be given to him. But let him ask in faith, with no doubting, for he who doubts is like a wave of the sea driven and tossed by the wind" (James 1:5-6).

So, we all have access to godly wisdom. Once received, we cannot stop there and just be content with wisdom. Psalm 1 says how we are to meditate on the law (God's word) both day and night so that we will have a life that overflows with fruit. This law is written on our hearts through our conscience and illuminated by the Holy Spirit.

Jesus developed and spread God's wisdom in a way that we probably wouldn't. He chose unlearned men as the founding disciples of bringing the message of Christ to the people. He chose them because they had a child-like faith to receive the message that he is the Christ. To be a little child is good because children receive things very easily. What did Jesus do with the disciples while they were with him? He taught them about himself because just as every child is to grow into complete adults, the disciples' destiny was to mature into complete likeness of Christ.

Let's see how Jesus taught them. On the day he rose from the dead, Jesus came upon two apostles walking along the road to Emmaus:"Beginning at Moses and all the Prophets, He expounded to them in all the Scriptures the things concerning Himself. . . .And they said to one another, 'Did not our heart burn within us while He talked with us on the road, and while He opened the Scriptures to us?'" (Luke 24:27, 32).

Jesus used the Scriptures to teach others about his divinity; it is a clear message to the disciples that the Scriptures must be studied.

The Lord Jesus, while being tempted by the devil, also used the Scripture to win victory over the temptation that came through the devil (Luke 4:1-13). If Jesus, the Son of God, used Scripture to rid himself of the devil, shouldn't we also nurse upon the written Word and ask the Holy Spirit, who illuminates the Word, to make it alive to each of us individually?

The Pharisees studied the Word and yet Jesus told them, "You search the Scriptures, for in them you think you have eternal life; and these are they which testify of Me. But you are not willing to come to Me that you may have life. I do not receive honor from men. But I know you, that you do not have the love of God in you" (John 5:39-42).

Jesus was saying here that even if we know the Bible cognitively, without faith and love, it profits nothing. Does this passage suggest the Scriptures are something to be taken lightly? Not at all. Jesus said the Scriptures testify about him. But the religious elite could not see him in the Scriptures because they did not love God. They loved their status as leaders. "Nevertheless,the solid foundation of God stands, having this seal: 'The Lord knows those who are His,' and, 'Let everyone who names the name of Christ depart from iniquity'" (2 Timothy 2:19).

Who do you see in the Scriptures?

★ 21 Dec 2008

I passed out at the gym a couple days ago. The doctors and technicians say that I was dehydrated; however, I know this isn't the case. The fact is, they love giving IVs. The technician even told me he was happy to give it because he doesn't often get to do it. I also overheard the doctor and technician speak about my EKG readings. They concluded it wasn't dehydration, but since the IV was started and they really didn't know what the cause was at the time, they went with the dehydration thing.

Well, you may wonder, if not dehydration, what is it? I don't know, but I think it's some chemical reaction with something I ate or it may be a deeper reason. I feel dizzy writing this. Life is full of situations that come whether or not you are eating right or wrong.

Despite the unnecessary IV, I reached another full year of life on earth. It is at this time that I wonder about and analyze the past year, how I thought it went, the things that I had hoped to accomplish and didn't.

I had hoped to become an officer in the military by this time. I had hoped to be an author, too. What I did accomplish was becoming a leader of the ministry I attended and I got accepted into the seminary. I rejoice on the things that I accomplished and the things that I desire to still accomplish. At the same time, I wonder why I have let so many vain imaginations plaster the walls of my mind.

As I age, life seems so circular. Though I try to break the cycle, I find myself in another circle moving along its own circle. I continue rotating until the grave will one day meet me.

Because life, at best, seems most vain, what should I spend these years doing? Many writings have been published on life's purpose; but the Bible says, "Let us hear the conclusion of the whole matter: Fear God, and keep his commandments: for this is the whole duty of man" (Ecclesiastes 12:13, KJV).

Our purpose and our wiring have been pre-determined. We war and fight over something that has been determined before our existence. As disturbing as this may sound, Judas Iscariot planned to betray Jesus. But this was not news to Jesus. It was recorded in the Old Testament many years before Judas received thirty silver coins for handing Jesus over with a kiss. When Judas betrayed him, Jesus said he did it "so the Scripture might be fulfilled."

You may still say that Judas or any man or women chooses or creates their own destiny; but consider this: If I take two runners and start one a meter behind the finish line, who is in a wheelchair and has strong arms, and I start an able and fit runner fifty meters behind the finish line, and the gun fires, who do you think will win the race? Clearly the race has been rigged to fit my point.

We all come from different backgrounds with different mental capacities, all of which are not chosen by us. One man smokes all his life and never gets cancer; another smokes once and gets cancer. One person is raised in a nurturing home and although he is schizophrenic becomes a successful doctor. A man without any mental incapacities, grows up in an abusive home, and becomes a hobo. Another has no mental incapacities, grows up in a nurturing home, and commits suicide. Do you think this doesn't happen?

There are over a billion people on earth and situations that span a trillion variations. Some add up and some do not. Why ramble on about this? Because life in its complexity can be simplified to understanding that the purpose in our life is to fear God and keep his commandments.

The problem for many is trying to understand God. His thoughts are not our thoughts and his thoughts are higher than our thoughts. We sometimes forget this because in our relationship with God, he can seem, at times, to be one of us. He is God and we are human beings and we need to understand that. He loves his creation and alone understands it.

Job thought he had sufficient knowledge about God, was righteous before God, and believed that he was wronged by God for allowing the devil to invade his life. What happened at the end of the story? The Lord challenges Job with questions. As the Lord question's Job's ability to save himself, he grew smaller in his own eyes and understood that his knowledge and power were of great limit, Job knew, having been wealthy, that his wealth was no shield against an almighty God.

It's God's wisdom that causes the world to consist in its complication of stars, moons, and planets. What man can honestly say he understands a fraction of what happens on earth? It is arrogance and pride when a wise man thinks to know it. The one truth that links all that appears to be chaos into purposeful order is to fear God and keep his commands.

Whether I am in the high time of life or in the valley because of a fallen comrade, I must remember to fear God and keep his commandments. Solomon said that even though a man may commit many sins for a long time and go unpunished, it is better for one to fear God.

So what does wisdom say for a soldier like me who spends his years on this planet with peril, success, love, and hate? Wisdom yells, "Fear God and keep his commandments."

It's here in the forty-degree, sometimes wet, December desert that I write this.

I've been working eighty-four days straight without a single day off.

Rockets attack daily.

But the conclusion is the same.

Fear God and keep his commandments.

★ 25 Dec 2008

It's Christmas.

My work schedule is the same. The dining hours are the same. And the postal exchange (PX) is still open. While much appears to be the same as any other day, it really isn't the same. At the post office, thirty pallets of gifts and packages of all kinds flooded the station—so much so that the post office asked for volunteers.

Among the piles, I have found packages for my very own. What a joy family is. In the midst of the wickedness and evil that abounds in many parts of the world, I can enjoy family. Though I miss their presence, whenever I hear their laughter and what sounds like joy on the phone, I smile.

I walked to the dining facility and had another opportunity to eat a good meal: chicken, stuffing, mashed potatoes, green beans, and a roll with orange soda to drink it down. These are the pleasures of life that make life worth living.

This day, however, doesn't end without a shot in the arm. One of my long-time friends of fifteen years saw his cousin dead on the floor. She was twenty-two years old. His Christmas will never be the same. In moments like these, I can only be still and listen for the silence to be broken by the one who has been wounded—and then without words, a gentle hug does wonders.

★ 26 Dec 2008

It is with despair that my friend now carries a burden of visual images of a lifeless cousin. I cannot say I know how he feels for I dare not quote Scripture, at least not now. This time is a time of quiet grieving and listening. Similar to the wise young man who counseled Job after time passes.

Job lost his sons and daughters in one day and his livelihood, too. It was seven days before his three friends spoke to him. It was longer for a fourth person, a young man who counseled Job by the Spirit of God. Elihu was his name. After talking to Job,his wrath against God was kindled even more because Job justified himself rather than God.

It's here that death can make more sense. It is here that we come face to face with God concerning the trials that exist in our life. Often I take the righteous route when facing adversity as if I am somehow above reproach. Jesus Christ was crucified for a crime he did not commit. People lied about him to have him crucified.

We often forget this when facing our own dilemmas.

Although we may commit fewer sins than before and although we are forgiven because of Jesus Christ's death and resurrection, we must never forget that God is smarter than us and when bad things happen, especially when we are doing all that can be done, we persevere through the trials in exchange for crowns that will never fade away.

★ 27 Dec 2008

A word from my Father encourages me:"Son, this journal writing will be blessed." It brings me great joy to have a father's blessing. It's amazing how powerful a word from a father can be. I often wonder how some who have broken relationships with their fathers, perhaps deeply wounded by their words and actions, live life happily. I only know the opposite.

My father raised me in the church; and in the quiet time of the week, he held Bible study in our house, which was mandatory for me to attend. Even though I went with them to church every Sunday morning and Wednesday night and attended the Bible study at home, those activities didn't reveal my parents' true values nearly as much as watching them live a godly lifestyle. Watching them pray in a quiet place was such a powerful image. It's these quiet times in prayer and study that I cherish the most.

While it's true that you can bring a camel to the water but can't make him drink, this isn't talking about bringing children to church or raising them in the admonition of the Lord. The wisdom of Proverbs 22:6 says, "Train a child in the way he should go, and when he is old he will not turn from it." This kind of training requires spending time with our children—time in which they see us pray to the Lord. If that isn't happening, we only have chance on our side as to how their souls develop.

I hear the struggles of military moms who have husbands that are stay-at-home dads and wonder: *do you care about the baby you leave behind?* In the desert, there are many job opportunities for civilians to join hand-in-hand with the military. The tours leave couples seeing each other only three to six months out of a year, year after year.

One year I was separated for eight months. That was difficult enough, but to do that year after year, like some of my co-workers, I could not willfully sign-on to that. They may say, "I have to do this job." Well, we all make life choices, but our children need time to be nurtured or else they will become emotionally broken and have difficulty being relational. And although we may spend time with them, if it is not in Christ, we will raise children who perhaps one day will have a 401K plan, decent house, and reliable car but this will all perish with them.

The truth of the matter is that many things lead to vanity. There is only one thing that leads to life—his name is Jesus. When Jesus becomes the focus of our relationship with our kids, vanity will be removed and we will come into a place of purpose and eternity.

The wise man looks ahead and sees danger and hides.

Today, I choose to hide in Christ.

★ 28 Dec 2008

Rocket attacks continue and bombing by Israel has caused the death of over two hundred Palestinians and four hundred wounded. Military and civilians have become casualties of war. So much fighting and hurt. It would seem a simple solution would be to just stop. What is it that causes this war and fighting for things we desire and cannot obtain?

In the past month, I have attended five ramp ceremonies—three for American soldiers and a couple for Canadian soldiers. Today, another comrade has fallen. He pledged to defend the camp and to defend a nation that helps to defend many nations in the world. The ceremonies don't get easier to attend. Sometimes they get harder. Standing in formation tears often form as the chaplain recites a familiar "the Lord is my Shepherd." Soldiers from many nations line up in formation to honor the fallen. Civilians off the flight line look on. Soldiers carry the coffin up the ramp and into the plane's belly. The bagpipes play "Amazing Grace." The plane doors shut. Taps plays. Tears drop. Soldiers return to duty in silence.

How am I supposed to live in the desert with peril all around me and a God who has no limit to his power? I conclude I am to live by the Holy Spirit by which I am sealed until the day of redemption. I am to pray, believing that God's plan will be done on earth as it is in heaven. I am to love those in whom I come in contact with and to present to them the precious good news that Jesus is alive and well and is calling for his bride, his church.

Until then, God has sent me to call out the one who has fallen into the world's enticements and cannot seem to get up. Likewise, he has sent the whole body of believers in Jesus to call on the nations of the world, including their own, to stand without doubting, knowing we have no permanently peaceful city today, but we seek one to come.

Believing in God has nothing to do with his ability to remove this war or to give us understanding of why he chose to take us through this war; believing in God has to do with a willingness to trust him. That means we are not to lean on our understanding but God's.

In the same way, a base security guard (in his own wisdom) might think that thorough checks every day of the same foreign personnel should not be done. But our superiors know, through experience, that enemies have used this kind of complacency in the past for military base destruction.

God's experience and his wisdom knows best, too, so we should trust God and always follow his instruction (not our own fatigue)in carrying out our spiritual duties or else we'll encounter negative consequences—maybe even destruction.

Trust in God is giving up on our ability to have every *i* dotted and *t* crossed. It's the ability to freely give God all things, trusting that in the end—which is most important—all things will work together for our good.

It is a blessed thing to be held in the hands of God and know with certainty that though we fall into temptations and sin, his grace is available for restoration.

★ 30 Dec 2008

I am still in the desert and it's still cold. I often feel like Jesus walking the dusty roads of Damascus. It has rained recently, so you can imagine how nasty everything is. I walk to and from work through puddles of muddy sand patches. Tracking this dirt into our mod housing never ceases. You cannot truly appreciate a concrete sidewalk until you walk on dirt every day.

I do enjoy going to the gym. Pizza Hut and Burger King are authentic to my US taste buds. The holidays are about giving, so we eat well and abundantly those days. Every other day, the food in the dining facility is the least bit tasty. Similar to drinking hard and soft water from differing taps in different states, the food here tastes somehow different. Some days I eat eggs with Chinese letters embroidered on them and other days french toast. These eggs do not differ in letters written on them alone, they taste different from the eggs I eat in Texas and I'm not sure why.

The cool thing about being away is receiving the letters and care packages. Every one I open is like revealing a little piece of heaven. They are priceless, like receiving gold bars from the US Treasury. I realize time needed to be spent in preparing them and that means someone really cares. Knowing that is so important for helping to keep the soldiers in the best state of mind. We work several hours a week without day breaks and grow weary without receiving these gifts every now and then.

The internet is lacking serious bandwidth. I have been warped back to the time of dial-up. Ugh! The constant disconnects makes chatting with my family rather annoying. In the bad and good, though, Jesus is always the answer. It's him who I depend on and some have taken notice.

My roommate's girlfriend has been asking me questions about Christianity, and we even got into a discussion about the difference between Mormons and Jehovah Witnesses. I told her that the primary difference between Christianity and every other religion, including Jehovah Witnesses and Mormons, is that Christ is the cornerstone. He is the foundation for which everything else is built upon. Remove Christ and you might as well remove the Ten Commandments and everything else because Jesus Christ, the Son of God, makes possible the redemption of our souls; that means that our sins are cleansed by Jesus forever.

I don't know if she got it; but upon leaving, she asked if I wanted some pizza and said she was buying. I quickly turned her down, but received a conviction about it after she left. I realized I missed the opportunity to invite her to have a personal relationship with Jesus. At that moment, I lacked the discernment I needed to effectively witness. She was sincerely curious about Scripture, not confrontational or combative in any way. I must keep in mind that people in this spiritual state need to be graciously presented Jesus Christ and helped onto the next step of tasting and seeing how good he really is. I pray the Lord will send someone else to her to do this.

★ 31 Dec 2008

A sacrifice is an act that causes discomfort in obedience to something higher. It may be for a carnal purpose, like working out, or an eternal purpose, like glorifying God forever.

Regardless, it is difficult to change the worldly desires that have formed in me. I have the knowledge that Christ is the greatest thing that has happened to me, but I find other things pulling me. I have a thirst to be something—perhaps a male bodybuilder or a salsa dancer. They may seem like very different aspirations,but both require a discipline that not all can or desire to commit to. Yet, I realize that aspiring to be like Christ is far better. Paul said that exercise helps a little, but exercising myself unto godliness is much better.

To be godly is to be interested in what God is interested in. God is most interested in the converting of souls. His last message to the disciples was to preach the gospel of Jesus Christ. Jesus gave them the power to heal and prophesy and the knowledge to set up administrations. He sent them to establish his church, and we are to continue their mission until Jesus returns.

Our jobs as those who believe and know Jesus Christ is the same all the time: fear God and keep his commands. His commands are to love him first, and then love his created man. This requires a sacrifice on my part. I must put aside my personal desires and agenda. It's easy for me to work out all day,but sacrifice is the burning of flesh, the crucifying of the will that fights to be in charge and not be submitted under the hand of God.

As I fight today to sacrifice my will to his, I recall that this is the day the Lord has made; I will rejoice and be glad in it. I will bless the Lord at all times and his praise shall continually be in my mouth. I will acknowledge Jesus as the only true God and savior. I will sacrifice my will to his.

January

★ 1 Jan 2009

God did not make Adam to search for porn.

God created male and female and blessed them. Their mind was good, void of perversion and confusion. Adam knew the voice of God; for before woman came, Adam spent time alone with him—so much time that the animals were already named by the time that God said, "It is not good for man to be alone." Eve knew God's voice, too, for he created her. But they both decided to listen to another voice—a voice that was not from the Father, but a foreign voice.

I, too, have run into a foreign voice that appears elusive, never talked about but simply stumbled upon by many. The internet allows for secret searches—at least they are assumed secret, so the voyage continues. With over a billion topics to look up, sexual images and sexual content is the number one searched item on the web.

If the subject is so vibrant, why does the church tip-toe around it instead of exposing it? Why is it behind closed doors? What of nocturnal emissions? I shall delve further into this peculiar topic; but at the moment, I need to discuss something from my youth.

My family was rather traditional. We ate at the dinner table together and had a full course meal: chicken, rice, and some sort of vegetable with cornbread. Sometimes, we had ice cream, popcorn, or pound cake for dessert. These were good times. One day I overheard my aunt and mother speaking about a douche. I didn't know this word, so I asked about it at the dinner table.

"Don't use that word," my mother said.

"Why not? What does it mean?" I asked.

She waved her finger at me: "*Don't use that word!*"

As a youngster shaped in iniquity, I decided to say the word again . . . and again . . . and again. My mother with angry-shaped brows and a deep, fatherly tone said, "If you don't stop, I'm gonna give you a retro-douche!"

At that point I stopped.

The topic, although not appropriate at the dinner table, gives a glimpse of the discomfort of speaking about female and male parts in my family, even in a medical way. My parents probably thought that not discussing it would make it all go away. I agree that discussing some sexual topics can cause more problematic issues, but to completely ignore them is unbiblical.

The twentieth chapter of Leviticus mentions several sexual sins that required death. This is just one chapter in the Bible concerning relationships between man and women, even animals. God covered every perverse thing that was being done at the time. Keep in mind this is Old Testament law and as believers in Jesus Christ, we are no longer under the law. With that said, these acts are still sinful; if a person has not received Jesus, he will be punished for them.

As we come closer to the coming of Christ, all these sins will be more and more prevalent because the church is ashamed to bring to light the sin that is being committed within their own houses. They fear being rejected by society or being labeled intolerant.

The truth is, though, that our airways are being flooded with acceptance, while our moral fibers decay day after day. I realize this must happen before Christ returns but, as a Christian in America, I do not think it has to happen at its present rate. Judgment is for God, but presenting the truth is what Jesus asked the disciples to do. Parents need to present the truth—the whole truth—and be bold in speaking the whole truth. Avoiding these issues is irresponsible given the nature of our easy-access to the images and content.

Returning to nocturnal emissions, last night I had one. These sexual orgasms at night happen in the daytime, too, and more frequently with men but also with women.

Until now, this has been a secret journey for me as it is for most people because the subjects of orgasms in dreams and masturbation are taboo, yet everyone on earth is affected by them directly or indirectly. It affects me because during times of abstinence, as I journey through the desert away from my wife,certain sexual dreams occur.

I considered these dreams to be evil and something that I could remove. So, I took time away from watching television. Other times, I prayed and fasted—three, four, or even five days without food and water. (I would not suggest going without water for more than three days, but I was zealous.) During this time, I read the Bible, chapter after chapter. I loved the Lord more and began to abhor sin more—not so much the sin in others as to judge someone who has not met Jesus but my own weaknesses.

Even so, the dreams remained.

I now believe them to be a biological necessity, particularly for the single person and those married away from their spouse. Of course, children have them, too.

I would not have come to this conclusion had I not taken all of the biblical abstinences. I prayed, fasted, did not put anything before my eyes that would cause me to yield to temptation; and yet, the inevitable occurred—I dreamt. The Bible says that when a man is tempted, it is of his own lust but that God would not give us a temptation that we could not handle. I mention this because what is brought into the light can be made useful for the cleansing of others.

A creature from a European legend called succubus presents another possible source of some of my dreams. I would not mention it, however, to have a more complete view of my challenge in the desert, I must. Succubus is a demon who takes the form of a beautiful woman to seduce men. I have had dreams with a visitation of a beautiful woman who approached me. When she came, an evil presence was in her. The moment I entertained her, I was overtaken. Not all nocturnal emissions come from intimate dreams, but some do.

The wisdom in Proverbs 5 mentions that a forbidden woman drips like honey and is more desirable than the sweetest things known to man, yet her path leads to a hell she is unaware of and following her will cause suffering. When tempted, call on the name of Jesus. Do not entertain it; for in doing so, you have already failed.

I interject this European legend with the Bible not to provide mysticism, but to provide a broad context to the subject of orgasmic dreams. Of course, a succubus is a female demon in legend, but the idea of her actual presence is as real as Lucifer, the devil, who comes to us as an angel of light.

Another prevalent problem that I need to speak about with truth is pornography, more-so because of its camouflage within the home. The military is quite aware that pornography is linked to the declination of one's ability to restrain oneself. They are also aware that with overexposure, a man is more inclined to attack and rape a female comrade in the deployment area. Even so, pornography is sold in the military base stores in the states and in the nations that surround me. Soldiers often try to bring it back here to their deployment when they return from rest and relaxation (R&R).

Before soldiers go on R&R,I brief them about this. After the briefing, men (sometimes Christian men) often ask me how thoroughly military customs checks laptops for pornography. Would a Christian man or women really ask this to another in the faith, even expose their desire to prepare for the cover up? On the other hand, these same Christians could just as easily quote some Scripture denouncing pornography while being bound to it. Then, you might have two bound people quoting Scripture to one another and nobody gets fixed.

The Bible says to confess your faults to one another so that you may be healed. Healing takes place when the shades are removed and truth sets a person free. This is the gospel and this is where Jesus' burden is light. We carry loads around that need to be confessed. Instead of confessing, we quietly pray to ourselves as if the gospel is void of relationship with others.

Confess and press and be free and clear.

★ 2 Jan 2009

Liberation is the benefit of trust, understanding, and truth. That's my proverb for this entry.

My wife told me today that some cute guy approached her with interest. Notice, he was *cute*. Some men would get angry with this kind of honesty from their spouse, but this anger only stifles the communication and places the relationship into the lie zone.

Just a reminder, Satan lives in the lie zone.

Isn't it amazing how so many cute marriages with two cute people end in divorce? Why? They forgot to be friends and enjoy conversation without fear of reprisal. They believe that when they got married to "the one" all other women or men should lose their appeal. This false pretense and lack of truth between them causes married people to act on their attraction and seek external relationships. Marriages that make it to thirty or forty years and can still say, "I love marriage" are marriages based upon truth.

I remember my mother talking about how gorgeous Sean Connery was and my father liked Toni Braxton. Neither one were disturbed by this. You may say, "These are television figures. What is the likelihood of these charismatic celebrities becoming a threat to their marriage?"

The point is they are people of the opposite sex to whom both were attracted. I am not saying that by observing someone's physical beauty, my parents are committing adultery of the heart. I don't care how many fasts you commit to or all-night prayer vigils you attend. You will still find others attractive. I know this because I fast and hold vigils and still find some women attractive.

What's the solution?

The Bible says we should leave the father and mother and cleave unto the spouse. This cleaving happens when the intimacy between spouses is no–holds-barred naked love without the trappings and entanglements that develop through lies.

★ 3 Jan 2009

I have come to realize that sleeping longer than eight hours ignites vivid dreams. Dreams of grandeur where I heal people in the name of Jesus or simply chased by a mob of angry people. This doesn't nullify the fact that medication and supplements may also induce similar experiences, but it seems apparent that the brain will get its need for activity fulfilled even if I do not choose to get out of bed.

The Bible says that with much activity come active dreams (Ecclesiastes 5:3). For instance, thinking about a particular concern will often cause one to dream about the concern. But we need to tread lightly in the area of interpreting dreams. Even though the book of Acts records that old men dream dreams, there is a type of dream not from God.

In the Bible, one of the clues that the Lord is speaking in a dream is that the dreamer isn't thinking about the same thought that appeared in the God-given dream. This was true for the wise men, too. After visiting the Christ child,they received a dream to go another way home. That dream came from out of the blue. They were not meditating on alternative means on how to get home; in fact, if it weren't for the dream, they would have gone back to visit King Herod.

Likewise, Peter had a dream to dispel his bias toward the Gentile people. The thought of Gentiles receiving the same exact blessing of the Jews wasn't in his thoughts, but his dreams informed and produced fruit for God's elect.

The day is coming, and in many respects is already here, with prophets and prophetesses giving prophecies in dreams of grand proportions to people who are living in blatant sin.

Today, dreaming is big business. It's taught in business classes and self-help books whose teaching philosophy is *dream big and live large for this thought will surely increase your way.* Some modern churches use the story of Joseph—how his dream saved a nation from starvation—to build their ministry upon a foundation of dreaming.

Teachers retell this story with much theatrics as the ideal situation. They promote the idea of gathering one's thoughts to define their big dream, which often consists of the dreamer's glorification. That's why people are willing to go to prison to get a reality TV show or a good book deal—for their own glorification.

While the idea of having a dream is positive, we must not lose focus of our real dream and mission. As believers, our dream is living life with the constant knowledge that heaven is our home and we are simply passing through. God's promise is for us to receive our reward in heaven. The afflictions on earth are light because for believers in Jesus they will end.

Joseph didn't receive his dream after discussing it with his brothers or hearing it preached from the pulpit. The dream was not a work of labor and statistical study of strengths and weaknesses. It was a personal, God-given dream. As life carried on, God moved and the dreamed appeared. Not a forced vision, but a natural flow of a supernatural activity.

I receive encouragement in the desert by allowing the natural process of God's supernatural actions to take place as he pleases. For instance, I have a dream to write this book and pray that its contents enlighten you. Is this dream given to me by God? I don't know. But I do know that all things work together for the good of those who love God and are called according to his purpose because it says so in Romans 8:28.

★ 5 Jan 2009

Do you recall watching the first *Rocky* film? Do you remember how you felt hearing that now famous song play while Rocky took his triumphant run up the steps of the Philadelphia Museum of Art? Did it inspire you? Make you feel like you were a part of his victory?

Music accompanied great triumphs in the Old Testament, too. David played the harp to get Saul's demon to leave and, afterward, to bring comfort to Saul. Jehosaphat and his army went to battle with the praise team and did not lift one finger in the fight. God took care of the enemy; all they had to do was collect the wealth.

As soon as I step outside to begin my walk to work, I pray and sing to the Lord. It may be a song I remember from Sunday school or something made up on the spot. I sing because the Lord inhabits the praises of his people. This thought is provoked by Psalm 22 in which the Israelites give praise to God as he inhabits them. The idea is to be fully submerged in the power of God just as Ephesians 5:18-20 (KJV) says, "And be not drunk with wine, wherein is excess; but be filled with the Spirit; Speaking to yourselves in psalms and hymns and spiritual songs, singing and making melody in your heart to the Lord; Giving thanks always for all things unto God and the Father in the name of our Lord Jesus Christ."

By nature, I am an introvert and, today, I am required once again to brief about 30-40 soldiers going on R&R. Without his power within me, fear takes root, which accompanies doubt and physical manifestations of sweating, stuttering, and lack of eye contact. My body takes over. My remedy is to sing praises to the Lord that stir up the gifts of his presence and power so that I can accomplish the task

Before beginning a fearful or heavy task, choose to magnify the Lord. Exalt his name. Glorify him. Call upon the Lord and he will show you great and mighty things you know not. You can quote Scripture, but it's another thing when you begin to sing it.

Let's have a renewed sense of the power of praise and worship and be like the four living creatures in the book of Revelation 4:8 who never stopped praising: "Holy, holy, holy is the Lord God Almighty, who was, and is, and is to come."

★ 6 Jan 2009

I must confess. Today, I sinned against the Lord of glory. I'm not saying that I have escaped all sin up until this point. In fact, 1 John 1:8-9 (KJV) reads, "If we say that we have no sin, we deceive ourselves, and the truth is not in us. If we confess our sins, He is faithful and just to forgive us *our* sins and to cleanse us from all unrighteousness." Today, I confess it.

Sin is the leading cause of death to mankind. Sin is the reason I am here in the desert. Sin is the reason I am armed with a rifle and enough ammunition to take down a plane. Sin is the reason we lock our doors at home and in the car and purchase expensive alarm systems. The works of the flesh manifest itself in selfishness, lust, greed, and many other desires that cause humanity to war with itself.

Scientists are actually trying to develop a chemical or a procedure that will help to prevent a person from committing these kinds of sins. Is it really possible to cease sinning? Is the cleansing process of Christ also available through our own development and research dollars in pill form? Can a pill cure us from the sin that is unto death?

It is my understanding that the sin unto death is unbelief; those who do not believe in Jesus and receive the Holy Spirit will enter hell. This happens because our soul is like a court document. In order for it to carry weight, there must be a seal; otherwise, it is an ordinary piece of paper without any power.

When we believe in Jesus Christ as the scapegoat for our sins, we receive the seal that gives our soul the power to be saved from eternal separation from God. This seal is the Holy Spirit.

God is goodness and love, so what's left in the absence of God is eternal damnation and hatred. Anyone who has ever been under the tyranny of hatred knows that it is not a good and loving place, but a place of sorrow and suffering. The Jews under Hitler faced his hatred. Soldiers who were supposed to protect citizens were gassing Jews in concentration camps. Nurses who were supposed to protect life were murdering Jews in hospitals. This isn't unlike the present day stress of Hamas; this is not love but hatred.

God has experienced hatred against him, too, yet he displays his love through granting mercy on millions who hate him. He provides wisdom for Hollywood actors and producers, for corporate CEOs and sports heroes to make millions. Most of them do not praise God but, instead, they make jest of him and his followers.

I have fallen today, but I am going to get up by asking Jesus Christ for forgiveness and pressing toward the mark of the high calling of God. Living in denial of my sin will get me killed because the wages of sin is death. Remembering the redeemer and pressing on toward his redemption through the power of the Holy Spirit is the only medicine I can take to cure my sin problem and lead me into life everlasting.

★ 19 Jan 2009

It has been thirteen days since my last entry and not without reason. The days past have provided me with a challenge that begs resonance. I am being challenged with primitive thoughts and feelings that have arisen due to a troop of mine getting in my face with a foul mouth that requires more than the natural mint.

I'd like to rant and rave about how this individual makes me regret my military choice. The one thing she has accomplished is for me to be sad like herself. Sometimes the most difficult people are bitter and broken and make life, if they can, bitter and broken for those whom they come in contact with. There is a remedy, which requires taking the high road—a soft answer turns away wrath. At the present, this seems impossible and improbable due to my initial tendency to defend and conquer and ask questions later, especially if there is an audience.

However, with sincere gratitude, I am writing from the peace of R&R in the land called Qatar. This rest and relaxation program is most useful for us who are deployed. Having worked 111 days straight without one day off, it is refreshing. I still, though, can't help sometimes but ponder the trouble I shall return to when all this rest and relaxation is over. Instead of wasting away this valuable opportunity by worrying meaninglessly, I shall shelf the desire and recall the wisdom in Matthew 6:34, NLT version, that says, "So don't worry about tomorrow, for tomorrow will bring its own worries. Today's trouble is enough for today."

So I rest.

★ 27 Jan 2009

I have returned from R&R and do feel refreshed. I enjoyed a little shopping and being a good tourist: taking pictures and visiting special locations.

There is a season for every purpose—a time to weep and a time to laugh. The time to laugh is over and the weeping has begun. If you are a newly-saved babe in Christ, you might say, "How can you weep?" You are the child of God and God is the best thing since sliced bread?" I only say, wait. For you, it is a time to laugh. It is the beginning of a newfound relationship that often offers joy and excitement. The prayers you pray to God seem to be answered and revelation appears to be flowing from a spring.

I only say, wait.

You may still be asking, "What is the cause of your tears?" This day, it is the dread of returning to a soldier I supervise who is very difficult to deal with. Those who work for and/or supervise someone are familiar with the clash of relationships at work. Ecclesiastes 8:9 mentions, "All this I have seen, and applied my heart to every work that is done under the sun: There is a time in which one man rules over another to his own hurt."

Well, time to let the light of Jesus shine in the midst of an untoward generation. You may do all you can to be fair and to exemplify Christ, and still receive the devil's reward of heartache and strife.

★ 31 Jan 2009

In the desert, this day marks the end of an age. The month is almost over and the new month almost beginning.

The day rises and I shout victory, for the Lord has heard my weeping and has come through on my supplication. It has been four days since the last altercation with my female subordinate. She yelled in my face about being wronged concerning her job performance report. She used the male bathroom, and I had to write her up. There have been a few good days and then another situation arises, many which have repeated throughout our working together. All I can say is that she has been a true rebel.

I've been bending over backward to make and keep peace and it seems to be a futile procedure. I changed the work schedule around to give everyone more time off. We did work seven days straight for months. But I give God the glory for these light afflictions are but for a moment. They arise and allow us as Christians to understand the power of God more.

Jesus was taken to the wilderness by the Holy Spirit and tested by the devil. Jesus learned obedience by the things in which he suffered. Well, the day has come and now is over. I learn to endure, like a good soldier of Jesus Christ.

February

★ 13 Feb 2009

Many days have passed since my last entry and you will soon learn why.

More marines have fallen.

I can't begin to express or know the feeling of the spouses and parents of the lost military members. We all know about the appointment of death, but it doesn't make it easier, especially when life is cut short by mischief and not natural causes.

As written earlier, these times are best spent in silence, at least for a moment. One of the hardest things to do is sit in silence, especially when we seek God's face for an answer to our dilemma. There are times in quick distress when I call upon the Lord and in moments, a situation changes for my good. But often, the gestation period requires days and even months, depending on the request.

I believe the Lord does not esteem one above the other based upon any physical disposition or net worth; however, as recorded in Scripture, the Lord will have mercy on whom he will have mercy. The Bible is full of prayers requesting a believer to entertain God with continual requests and even to acknowledge him in all our ways. But I believe the progression of a believer growing in the grace and knowledge of God and in the power and demonstration of the Holy Spirit is not unlike growing in the knowledge of any discipline you may endeavor to engage in.

For instance, decisions to make rank in the military faster than anyone else is driven by certain parameters as time in grade, but when testing comes,I must prepare my mind to do well. God also promotes us based upon our calls. But when the testing of the Lord comes, which often happens at the start of a useful ministry, we must pass his test to receive our promotion. In obedience, be satisfied with where God has you because where you are is important to the overall mission.

Growing in the Holy Spirit requires time—time spent in prayer, praise, reading the word, fellowship, fasting, and meditation. Practicing these disciplines will open a believer's eyes in the Spirit. Those who live according to God's righteousness have a greater power with God. It says so in James 5:16: "The prayer of a righteous man is powerful and effective."

When Job's friends accused him of sinning, Job could have become very angry at them, but at the end of Job's test, God told him to pray for his friends or else they would be under a curse. Because Job was found righteous before God, his prayers for his friends protected them against the wrath of God.

There is a righteousness that we receive through the blood of Jesus that allows us to enter into heaven. This righteousness is placed upon one who confesses that Jesus Christ is Lord. The eternal righteousness of God substituted for our filthy-rag righteousness.

There is also a righteousness that a man chooses to live by after his conversion. This righteousness is a continual decision to abide by God's wisdom. For example, the Air Force commander has been tried and tested through military testing. After accomplishing successful missions, he is not without mistakes, but his mistakes are minimal. He receives awards and praise from his subordinates as those over him bestow upon him honor and privilege.

The Lord, likewise, bestows privilege on his saints who choose to accomplish more work (fasting, prayer, meditation, and reading the word) and pass more tests. This is not working for salvation. The salvation has already been completed at confession and conversion. This is growing in grace, power, and influence.

Remember, God's secret is with the righteous.

I have spent a good amount of hours in the past days running on the treadmill, doing different types of strength training, and reading workout material. Yet, I realize that no matter how much I read or how much knowledge I obtain, my physique will only change with hard work, sweat, and tears. The better the diet and exercise regime, the better the physique looks. The less I cheat, the quicker the results. And, of course, the same goes for the righteousness that abides in God's wisdom.

Well, thank God that he has mercy and loves me. However, don't mistake for a moment God's love and compassion for me as being uninterested in my behavior, for he is always interested. Sometimes, God just watches in silence to see what I will do in his apparent absence. It's the absent feeling that tests whether or not I love God and follow him anyway.

I press on in the midst of being in the desert.

★ 15 Feb 2009

I have acquired some more knowledge and learned a valuable lesson concerning burdens.

I was speaking with a longtime friend about being obedient to all the words in the Scripture. He accused me of living in idol worship with my zeal for hitting the weights and running on the treadmill. Of course, my flesh began to heat up. But in the midst of pondering his words, I realized that in some ways he is right. I must not forget my first love.

Things pertaining to this earth can and will try to take the place of God, whether it be less prayer time to have more workout time or less reading the Word in exchange for reading about supplements.

Anyway, he tells me that he is going to start to witness to random people on the streets and that he is reading his Bible every day, and has even begun to read it to his mother and sister.

I read aloud to him the Scripture that says, "If a man will not work, he shall not—"

The phone disconnected, so I called again and began another subject but soon returned to working and eating.

The phone disconnected again, right at the same spot. It then dawned on me that I am giving him an extra burden that at this time in his spiritual journey is not required. Sometimes, seasoned believers hold unbelievers and new believers to a standard of living that they themselves have reached. Unfortunately, looking at others through your own lenses often produces frustration and offense.

God hasn't come back yet because he is waiting for the precious fruit of the earth, which are believers in Christ, and he has long patience for them to come to him. As lovers of Jesus, may we also have this same endurance to wait for the precious fruit. He makes all things beautiful in his time.

This is the beauty of Jesus. I am amazed that sometimes when we just leave someone alone, how the Holy Spirit can begin to minister to that person. When they are ready, they will edify us with their newfound insights.

We must not give new believers a burden that God has not. In this regard, I will hold on to this one treasure: "My burden is light." This verse often flies above my head, but I think Jesus' burden is light because he, unlike the devil, gives only what I can bear and nothing else. If I am overwhelmed, it may be because the devil is giving me burdens that Jesus did not.

Jesus' burden will be enough, but never too much.

★ 16 Feb 2009

The day is hot and the night is cold. I guess that's typical in this desert. People think that it is always sunny and dry and hot in Afghanistan, but actually it rains a lot here and,yet, there is no vegetation. Instead, the ground sucks in the water and turns it into saturated pallets of mud. Walking in it and tracking it inside still never ceases.

I find today that when I would do good evil is present—knowing on the inside that when I have chosen to do the right thing, doing the wrong thing is still a present option. There are many ways to live that seem right. Those living an ascetic lifestyle may appear without question to be living above present evil,but are they? One of those apparently good ways to live is to fast. Although fasting may be performed for good, the presence of evil might prevent the desired result.

Because of this, I am battling in my mind whether I should return to this age-old discipline. Seldom used by today's Christian, the church at large places more emphasis on tithing because the method of tithing is much easier, requires less time, and builds bigger edifices than does fasting. When many Christians do fast, they formulate new kinds that the Bible never mentions, like a candy fast—but chips and soda are fine. It looks to me like nothing has been given up.

Traditional fasting involves either abstaining from all foods and drink except water, fasting from all food and drink, or fasting from pleasurable food, basically a green food and water diet. We also must go on a fast with a focus and purpose. Isaiah mentions the fast required by the Lord, one that looses heaven's burdens and gives bread to the needy. It's a fast from our selfishness so that the excess from our fast should be given to another. Fasting focuses attention on Christ for his empowerment to a crisis ahead.

Some people in the Bible did not struggle too much with fasting because what they desired from the Lord was such a passion for them that they didn't really want food. King David fasted for his child to not die, and Esther fasted for the deliverance of the Jewish people from annihilation. Biblical witnesses illustrate that a believer has more power fasting for God than one who refuses to fast. It's not the fast that's so difficult; it's the lack of passion for a cause. So I ask myself, "What cause do I have enough passion for to commit to fasting so that I may see the power of God on a greater scale?"

Today, I reflect and realize I need a passion for Jesus. I need to be as Jesus, who said that his will is to be of the Father's business. Always, Jesus was of the mind-set of the Father's business.

★ 17 Feb 2009

His name is Mercy.

God is merciful and God is love.

I sometimes can't understand why some people think God's character has changed from the Old Testament (wrath) to the New Testament (love and mercy).I find he is the same yesterday, today, and forever and that Solomon's words are true:"There is nothing new under the sun." There is no sin we commit today that was not committed in times past, and there is no new love we have today that was not already loved in times past.

Tears well up in my eyes at the disobedience of Israel knowing that I also am disobedient. Yet in the book of Judges, God is a perfect picture of mercy and love, and this book is found in the Old Testament. It was in the outcry of the Israelites that God sent a judge to deliver them from their famines and bondage and put them back on top of the blessing pile.

During the time of the judges' lives, Israel prospered when they worshiped God alone and obeyed his commands. But, unfortunately, not having nurtured a personal relationship with God, the people often served idols. They forgot about how their Lord and God said he would raise a nation to put them in bondage if they worshipped other gods. They chose to worship other gods and went into bondage. The process of turning back to God and receiving his blessings would again ensue.

Looks like mercy and love to me, just as is recorded in the New Testament, too.

How should we live in light of this story? We should make great effort to know him personally so that when one of our own deliverers passes away or goes on vacation,we don't change our theology and join a worship of idols like the Israelites did. These things happen. Great leaders of God care for and teach people the ways of God, Then they pass away and the congregation following him passes away as well. This should not be. The congregation should be equipped to go out and teach others.

Remember the mercy of God and his love, and let us worship him and him alone.

★ 18 Feb 2009

The battle rages on in the skies and in my soul.

I still fight to be of one mind with the Holy Spirit while the thoughts around me are not thoughts of purity, but vulgarity and filthy language abounds. This environment is not the best for trying to have a clean and free mind. But it is the best environment for the greatest amount of light to shine. The darker the environment, the greater the light appears.

While I love working with Christians and would love a utopia of on–fire, born-again believers in my circle, this does not help the unbeliever who is blind to God's truth. Believers are often scattered in different work environments, not too many in one section; and believe it or not, this goes for church jobs as well. Don't think that if you are a chaplain or chaplain assistant that you're environment is any different because it is not.

The light shines when our conversation is holy. When our conversation is that of the gospel of Jesus Christ at all times. Colossians 4:5-6 calls us to "walk in wisdom toward those who are outside, redeeming the time. Let your speech always be with grace, seasoned with salt, that you may know how you ought to answer each one." I am not speaking about a robotic, verse-quoting speech at every hour of the day. That isn't what Colossians 4:5-6 means. It means our conversation should be easily understood and peaceful, easy to be entreated. It means our conversation is void of gossip and belittling others, speaking rightly about a person who is sinning.

It has been hard for me not to keep my mouth off of others in the desert. My conversation with my wife after hours of talking sometimes is soured with negative remarks; and for this, I ask the Lord's forgiveness.

Who do I talk about? Well let me just say that there are two things concerning handling problem employees wisely that I am learning. The first is that if you have prayed for God to handle it, continuing to speak negatively about the person only hinders God from moving on your behalf. He will not be used like this. God has his way with the heart of any individual. If you follow the lifestyle God has asked us to follow, he will take care of the situation. So,"do not rejoice when your enemy falls, And do not let your heart be glad when he stumbles; Lest the Lord see it, and it displease Him, and He turn away His wrath from him" (Proverbs 24:17-18). In other words, God can easily restore the situation to its difficult status.

Today I am making a decision to let my lifestyle be that of Jesus Christ's Gospel and the words that exit my lips, which come from my heart, will not cancel the prayers I have prayed.

★ 20 Feb 2009

I am nearing the end of my journey through the desert.

With only forty days left, I feel this is a time to burn—a time to press through, fast, and wait and see the glory of God.

Something happens at the mere thought of change. I find a struggle within even to the point of sometimes physically shaking. Similar, I guess, to someone who has quit smoking and finds the body shaking as the tar is being cleansed from the system. The brain aches in defiance to the decision; and if submitted to the urge for nicotine, one would find temporary relief only to once again be brought under the bondage of horrifying pollutants destroying the body.

Sin and disregard for God's Word does the same thing. In the desert, the habit of watching episode after episode of defiance to God corrodes my thinking and hinders my ability to commune with him, to discover his secret.

What does it mean in Proverbs 3:32 that his secret is to be revealed to the righteous? How far or close is this secret, and will God reveal it to anyone? In Proverbs 15:29 it also says that his ear is open to the prayer of the righteous. Many think that if they just call upon God he will answer. If the situation changes, they automatically assume God answered their prayer. But how many have worried over something that didn't actually come to pass without a prayer uttered? I understand that blessings not only fall on those saved but also on those unsaved. However, there is a Word from God that believers receive that those unsaved (not having a personal relationship with Jesus) cannot hear or see, which isthe revelation of Jesus Christ.

Our relationship with God is difficult, if not impossible, to place upon paper. Imagine seeing the back of God as Moses did. If we are honest, it is impossible to do. We can read the story; but unless the experience of Moses has been personally experienced by us,we can only get a glimpse of the experience through the words of Moses. It is similar to trying to get an unsaved person excited about Jesus by telling him some of the profound things Jesus has done. But he has not seen or experienced Jesus personally. We can try to describe it, but it's not until he is enlightened by the Holy Spirit that he can enjoy and appreciate what Jesus Christ has done on the cross.

Yet another day has come and gone. I know there are things I need refined. I thank God that I have the time to still refine them.

★ 21 Feb 2009

Another poke has been made at me concerning the spiritualization of Scripture.

I am to understand by this man that the Garden of Eden and the trees are all spiritual and that the tree did not exist nor did the fruit that Adam and Eve ate. So without further waiting, let's read the Scripture in light of this new revelation:

So the spiritual serpent is more subtle than any spiritual beast in the spiritual field, which the Lord God had made. The woman is deceived by the spiritual serpent into eating the spiritual fruit who then hands the spiritual fruit over to her husband who also eats of the spiritual fruit. Having sinned and their eyes being opened, they sewed spiritual fig leaves together to make spiritual aprons. God, knowing of the sin, walks through the spiritual garden and curses man and woman and serpent. Then God clothes Adam and Eve with spiritual skins (paraphrased from chapter three of Genesis).

You may think this is humorous, but people often rewrite Scripture this way. What is the purpose? To minimize its authority and to maximize mystical interpretations. It is the perfect plan to remove the Word altogether. People say things, like "Scripture has been changed so many times and man has tampered so much with it, you cannot use the text in a literal manner." But if the foundation (the Bible) is removed, who shall be able to stand?

This reminds of me of when I spoke to my friend who had just purchased a BMW 325i. Another friend, in light of the high gas prices, advised him to use low-class gas. "I do the same thing in my Lexus and it runs fine," he said.

The manufacturer's instructions say to use high-class fuel only,but my friend, listening to his friend's opinion about a car he did not design or build,put low-class gas in it. Well, the engine started making a clattering sound after putting the fuel in the vehicle, and it cost him a lot of money to get it fixed. I only assume that if the instruction were spiritual, he may have been spared the event. But since he is living in the physical world, it became apparent that the situation required a physical solution. By following the physical rule of placing high gas in the vehicle, it now runs well.

What does this have to do with the Bible? Every religion and doctrine has been codified by a book. The US government has the Constitution. My job has operating instructions, and doctors have medical regulations. Why are these rules written? So that a doctor or government official or myself will be guided by the contents. If I feel the writing is wrong and choose my own path, I will face reprimand and may even be fired. I could also endanger lives. Why? Because I gave feeling precedence over written regulations.

Likewise, 2 Timothy 3:16-17 says that"all Scripture is given by inspiration of God, and is profitable for doctrine, for reproof, for correction, for instruction in righteousness, that the man of God may be complete, thoroughly equipped for every good work."

Notice the word "every" because it is important. It means nothing has been left out of Scripture for the man or woman of God to become full in righteousness. When a man or woman says that the Scripture is to be taken lightly and that its contents are so deep that it enters a spiritual realm and that it produces no earthly good, we have erred from the truth. The

Bible says we are to be wise and that we are to study Scripture to show ourselves approved by God.

I pray the Lord will open all of our understanding concerning the importance to remain within the bounds of Scripture and not to leave the standard regardless of a special revelation or dream by you or anyone else. When we leave Scripture, we enter a world where, at best, we all are amateurs. Perhaps that is why God selected very few people to carry forth his revelatory world changes: Moses represents the Old Testament law and Peter and Paul the New Testament church.

Many admonitions in the New Testament tell us to beware of dogs, workers of iniquities. Why? The enemy, as a roaring lion, seeks to devour us, and how else to do this than by undermining Scripture at every turn. There is spiritual power from God anointing a believer for a work, but it never contradicts what has been written. Remember Satan has power to deceive by using crafty methods, so judge all things through Scripture. God has a plan and his revelation is revealed when he desires it to be so.

★ 28 Feb 2009

As the day breathes upon the last month of my tour, I am glad to say I see the end arriving. What is in store for the final days, and what challenges await? I don't know, but excitement ensues as the time to leave grows closer, and the realization of having a wife in my arms is better than any dream.

I still cling to the thought that I do not worry about tomorrow for tomorrow may never come, but this day is all I need to concern myself with at this moment. Sometimes I do get caught up into spreading current problems into another day when the next day would have otherwise been without problems.

Today is not just the day that God has made but a day I am actually off work. This rarely happens in the desert, so I took the opportunity to do some needed house cleaning. I mailed some things home that I will not need for the coming month. My wife desires a necklace from Habib's jewelry store, and that will be my last purchase concerning souvenirs.

March

★ 01 Mar 2009

Working through the pages of Scripture trying to find the secret of living properly, I come across a verse concerning the meditation of man. Psalm 1:2-3 mentions that in order to be successful or prosper, one must meditate upon the Word. If I mediate upon the Word of God, it prospers me. Never forgetting that at the present distress, there is a season for prosperity.

People are amazing creatures that adapt to their environment. After months of running into the shelters at the threat of bombs and gaining strength from the weight of carried weapons and ammunition, we can become complacent. This is not true for me. Daily,I desire the freedom to enjoy my warm bed at home and the relationships I have left behind.

As the day ends, I realize this should be the final month. The time of departure is near. I need to make sure that everything is aligned and that my replacement has a smooth transition. It is my desire to be plain in my explanation of the job that I perform. Making sure all personnel are well and accounted for and authorized.

These last days are also a testing period as the human condition tends to be lax in anticipation of leaving. I will fight the temptation and work to maintain integrity. A believer is to be ready in season and out, always redeeming the time because the days are evil and the field is white, ready for harvest.

How do we get people to come into the harvest?

Be wise as serpents and gentle as doves.

★ 02 Mar 2009

My performance evaluation thus far has been sent, but my job is still not complete. I am working on a document to hand to the oncoming personnel as I have made changes to the operation and would like to present a document for reference. My supervisor at this time is stationed at another base in the desert, so our relationship has been one of phone conversation and email.

In the desert, God has blessed me with what I believe is a grand supervisor, one that is on guard as a shepherd for his sheep. If you ever have a supervisor that brings you turmoil, pray. For the Lord fights our battles; and in prayer, you will always have the victory.

I find in my life there is a season of good bosses and a season where I must endure the horrific. But I remember Proverbs 21:1, which says that the heart of the king is in the hand of God who is able to turn it however he chooses. This was true for David who confessed that the Lord had delivered him from all his fears(Psalm 34:4).

This day for me was a day of reflection. Appreciating my supervisor; but on another hand, dealing with a hardball that has managed to task me with part of his job. He is here with me as I write. If you have never been in this situation, you will if you work.

My motto for this day is to remember that trials and tribulations are appointed unto the believer, but also try to make sure the trial is because of righteousness and not sin.

★ 21 Mar 2009

I pray my journey thus far hasn't been too tiresome and that you may feel some reward for my story and trial. I pray that God impart life and discernment for your own journey to follow.

As you may notice, times of writing leap across days as the wheels of replacements and briefings are everyday business. I am gearing up for redeployment; things are exciting as well as painful because as the time approaches, time seems to slow down.

Over the past days, rockets have been launched into the camp and soldiers have died. The amount of injury that is endured in the war is hardly mentioned. Death in some cases, in my opinion, is the better option for some soldiers who are wounded. Many only get separated from the military to commit suicide later. Soldiers are burned, limbs are removed, and previous relationships are severed. It is a sad time.

The other day, the building shook from the impact of rockets exploding. It's amazing the power they wield. We also have to endure rising assaults on the sexual front. Briefing after briefing, by officers who are experts in the area of human behavior, try to stem the tide of this rising problem among our soldiers. The problem resting more in the heart and decisions of some soldiers rather than in the reporting. As made public, some reports have helped and others have been to no avail. The briefings do give us a glimpse, though, on how unseemly our generation is.

Does this require mentioning again? Am I beating the drum in the midst of so many other drummers that the sound I make is seldom heard, if at all? I hope so because "rebuke is more effective for a wise man, than a hundred blows on a fool" (Proverbs 17:10).

I do understand how easily the hearts and mind can fail. These past few days,I have endured sexual temptation like you wouldn't believe. The amount of pornographic and near pornographic images available is astounding. Please understand. The military prohibits pornography. It isn't sold in the PX nor is it allowed on personal computers. I brief troops on this three days a week who are going on R&R. But the nature of men's sexual drive and insatiable lust consumes their lifestyles. It could be blamed partly on a lack of females on post or the amount of time a person is away from their spouse or having a previous addiction to porn before arriving to theirarea of responsibility.

Somehow, x-rated movies and songs of debauchery pop up on shared files on iTunes, in addition to Christian movies and music. When logged into the internet, I have to navigate around the destructive and highly tempting stuff owned by other users just to get to God.

I recall my mother telling me how Lot was vexed night and day because of the sexual perversions he experienced in Sodom and Gomorrah. His two daughters, for example, had no problem getting their Father drunk and sleeping with him. Maybe I can understand one of them, but both? Another interesting part of the story, and something I haven't heard from anyone, is the reason Lot stayed there in the first place. Incest, sodomy, and all sorts of sexual sin happened in Sodom, acts that are banned in the Bible, but Lot stayed because of his wife who loved the present world.

Be careful not to fall victim to sin because of a spouse. God still considers this sin. Keeping our integrity, like Job, sets the believer up with blessings, and not curses, in the midst of a family who commits sin. Setting other gods (like pornography) before God is a matter of choice. We cannot stop the desire of the flesh to worship things instead of God, but we have the opportunity to cast them down as soon as they enter our mind.

The hardest thing about living continually subject to God's will is to not allow the door to crack, for this only allows the flood gates of more sin to open. You'll then find yourself in a downward spiral clinging onto false doctrine in pop-culture as well as psycho babble from those with degrees in humanism.

Christ is sufficient for the journey and is all that mankind requires to become complete.

★ 22 Mar 2009

I have been watching *Lie to Me*, a show based upon the technology of reading individuals through body language, recognizing facial expressions and physical gestures, even hand placements. I find the subject rather interesting, because I believe what they reveal about a person are more correct than false. But I must admit that any body language interpretation is based upon a particular culture since each has various norms to the way they bodily express themselves.

Even with these observable techniques of knowing people, God will allow a believer to know another person without studying body language. The Holy Spirit in us gives us discernment when we are in tune to the desires of the King of kings and Lord of lords.

This gift of discernment is for the edifying of the body of believers as well as to be used to attract the elect to Christ. We must have in mind always to be about the Father's business. It is his pleasure that we serve him and become lights that cause others to glorify him. Believers in Christ need to demonstrate the power that God gives in discernment to strategically witness. As you recall, I did not do this so well on 30 Dec.

A strategic method of witnessing is to allow peace to be the guide. I witnessed this way to a young, Arab cashier at the Green Bean once. I ordered a spiced apple cider and a banana nut muffin, which he recommended. It sure did go well together. I'd been there many times, but had never spoken to him much. This time,I asked him whether or not he had heard about Jesus.

"Yes. Jesus was a great prophet,"he said.

"Well, have you also heard that Jesus not only prophesied of his death and resurrection, it indeed came true? Really. To this day, Jesus' body has not been found because he was resurrected. He wants us to be resurrected, too."

His polite smile was decreasing just a tad, but I continued, "Look, when we believe in Jesus and his death and resurrection and ask Jesus for forgiveness of our evils, we'll be saved from the coming hell."

I could sense his sincerity in at least wanting to hear what I had to say.

"Hey, I never heard someone describe Jesus like that," he said. "I'll think about it."

I smiled. "Believe and put your trust in Jesus. He died to save you, too." I waved good-bye and left.

If I didn't feel a peace after first mentioning Jesus' name to him, I would have remained silent. When Jesus told his disciples not to cast pearls before swine(Matthew 7:6), he means do not give the precious wisdom of God to those who will only turn it around and try to disrespect you. Here in this foreign country, on that Green Bean visit, I relied on God's wisdom to know when to give the gospel to the cashier and when to remove myself at the proper time.

I know some people who would have given this young Arab a different kind of a gospel as a safe and peaceful way to communicate about Jesus. But 1 Thessalonians 5:3 says that when people cry peace and safety, then shall sudden destruction come upon them and they shall not escape. A portion of our generation will only endure good teaching for a moment. As time progresses, they will run to those who teach a gospel that is not from God. They will turn toward a gospel that removes Christ and holiness and replaces it with universalism—a doctrine that all roads lead to heaven and that Christ cannot be the only way.

The allurement of peace at any cost would seem to be God's way, but is not. Let us be careful to seek a unity that is under the umbrella of Christ alone. Any other umbrella reproduces the Tower of Babel, which to date we have not recovered from. The Tower of Babel in the Old Testament represented a time when one language was divided into many,and God dispersed the people who were unified but not unified under Christ.

Today, I have come to know just one more thing that the world is producing and that Christ already enables a believer to have—peace. Whether or not you have watched *Lie to Me,* we have Christ and he is sufficient.

★ 23 Mar 09

My replacement has arrived and training has begun.

My last days are soon approaching. I have tried to carefully give my new arrivals as much information as possible to make their transition smooth. They have a hard task ahead as things continually change. After my removal, things will morph and change and then change again. For the time being, I must give them the current information of how things are processed at this moment.

I find this passing of the baton similar to Paul giving Timothy detailed information on dealing with church order, with heretics,and the elderly. He tells Timothy to be bold in the Lord and in the power of his might. He tells him that he hasn't been given a spirit of fear but of love and of power and of a sound mind. Paul also tells Timothy to study, to show himself approved as a workmen that need not be ashamed, rightly dividing the word of truth (2 Timothy 2:15).

Some may wonder how I transitioned to Timothy in light of training my Army replacements. As believers, we need to always have the mind-set that Christ is in all and through all. We do all things to glorify him. The stories he gives us in Scripture can be applied to every area of life and trial we may face.

There may come a time when you have to transition from one job to the next. The way in which you transition will be remembered and the new arriving personnel, whether voluntary or involuntary, need to know you have poured out your best. This is my objective, and I pray that I shall deliver.

★ 27 Mar 09

I must admit that I feel an anointing over these words,as I believe they are not my words entirely and will do some good for someone.

Am I to believe the minister today who utters prophetic words of the year of the coming of Christ? Am I as a believer in Christ under a fog as to the time that is given to all by the

Spirit of God? Am I to take notice and wonder with amazement at the teaching that is being made about numerology?

What am I to believe about the time to come? It depends on what I know about God.

I have the desire to expel a myth, a generational curse that is upon many—a lack of knowledge by the believer.

Americans are amazed by low SAT scores and the low number of literate individuals who graduate from some of our high schools. Americans are amazed by the number of schools who still don't have computers or have systems so old that using it for anything other than a typing tutor is useless

But when it comes to learning about God, none of this matters. Many believers are wandering around sucking up teaching after teaching about God when they only need to open the Bible and allow the Holy Spirit to speak. When a believer leans on the principles of the world system for advice and direction, he has admitted a deficiency is his learning and reading of the word. "All Scripture is given by inspiration of God, and is profitable for doctrine, for reproof, for correction, for instruction in righteousness, that the man of God may be complete, thoroughly equipped for every good work" (2 Timothy 3:16-17).God has always desired a relationship with every believer personally. You need not have any man teach you anything, but the Holy Spirit teaches you all things and brings all things to your remembrance.

The school system and job training courses today have influenced the church to such a degree that we have resorted to using training tools that model after the world. Want a larger church? Create a church brand, like "Jumping for Jesus."Ask your pastor to jump over fire with his motorcycle in front of some crosses to make a point about the cross. Or create a club called"Hookers for Jesus," where Christian women dressed like hookers witness for Christ. *Transformation* is what Christ taught not *conformation.*

The Lord desires his people to cry unto him. He offers the sacrifice of personal praise, adoration, and worship. We are to be ready to tell people the truth in season and out of season and to pray without ceasing.

I received a prophesy on my way to the desert that I would learn something and be increased in knowledge of my Lord. I must say that learning has taken place. I come to know more deeply that Christ is sufficient and we need no other.

As believers, we must understand that Christ will not fail us. He will lead us and direct our path when we ask him to do so. Keep in mind that going to the desert is part of the process and necessary for us to discover ourselves. In the desert, our character is revealed. Here, a married couple learns whether the marriage is infidelity-proof or not. Here, financial dispositions are revealed. Here, church attendance and fellowship with believers is tested.

There is no lack with Christ. In the desert I have come to learn this more and more. So, what does a believer need to know concerning the last days and things to come? Only what the Bible says.

Be ready!

★ 28 Mar 09

Are you judging me?

How does it make you feel when someone asks you that? If asked to define "judge," would you use the dictionary? Would the dictionary even help without a context?

I've learned in the desert that "are you judging me?" is often a defense mechanism used to defer the guilt a person feels over their sin. What they are really saying is"you got me, and I have nothing else wherewith to defend myself. So try this, Christian!"

Here is the problem with the question: When I tell people that Jesus is the only way to heaven,I am not judging the other person. I'm giving them knowledge. The words are not my own but Christ's words. Clearly, when the judge smacks down the anvil and the police officer arrests the lawbreaker, it's not the police officer who has judged the man or women, but the judge.The officer is only putting the lawbreaker in jail.

It is the same for a believer who simply mentions Christ's words to God's lawbreaker. A problem is that the words of Christ could potentially be laced with tones of condemnation and come across as a personal character judgment rather than an eternal judgment.

We need to realize that there are several types of judgments in the world and not all are against God's will. The judgment pertaining to God's domain is final judgment. Am I to personally declare who is going to heaven or hell? No. I only preach the gospel. I mention that Jesus is the only way to heaven and no other way. People enter heaven or hell upon hearing what the Bible has already said, not my words, not my judgment, but God's. Then they need to make a decision to believe or not. Do I know who has or has not done this?

The Bible says you shall know who his disciples are by the fruit they bear. May you, as a believer, not fear the non-believers' question: "Are you judging me?" because a ready heart at the revelation of one's sin breaks in the forgiveness that Jesus offers rather than seeking a way to avoid guilt. If they play the blame game, abort. Don't allow the devil to use this scare tactic as an excuse to change your position.

In the desert, I have been asked, "Are you judging me?" from people of many nations and cultures. Amazing. Jesus is truly global.

Remember, there is much power in the name of Jesus and it is his name that will always render a response, whether positive or negative.

★ 30 Mar 09

The time is near to leave and I am ready.

The first flight left me. I was annoyed and sad, but I am scheduled for the next flight. I ended up staying in the PAX (passenger) terminal for twelve hours. Since I worked at the terminal, I stayed inside the Joint Personnel Reception Center (JPRC), watched television, and spoke with the new replacements, so the time didn't go by too slowly.

My emotions were moving back and forth as I pressed in to hear the Lord. I didn't know why the first flight left me or why my flight coordinator scheduled my departure so close to the time I needed to catch the next flight. It worked out, though, and I left with the army. The first flight, I would have been alone. On this flight, the Air Force Personnel Contingency Team (PERSCO) took personal care of me. I was driven in a truck to the depot to drop off equipment and given lodging at Manas. Had I taken the other flight, this probably would not have happened because of the time of day I would have arrived.

The flight back is more nerve-racking than coming here because I just want to get home. Delays make you just want to scream, or maybe it's just me. Determined to remain in the spirit of one professing Christ, I do like others. I wait quietly.

★ 31 Mar 09

The time approaches when I will finish my journey; and like Paul, I say within my mind that I have fought the good fight of faith; now what is ahead for me is proper rest.

Proper rest in the material world is peace in the middle of the storm. Paul was speaking about the rest a Christian receives at death. The peace I am referring to is the peace I hope to receive sleeping in my own bed with my wife. The satisfaction of a job complete and a sentence fulfilled. There is never such satisfaction and pleasure with one's life until after you have been tried and then released. This pleasure may only come from such a trying.

Jesus learned obedience by the things in which he suffered. His suffering produces our pleasure in being redeemed from the coming judgment on the unbelievers. Look toward the satisfaction. Look toward the pleasure. Look toward the fulfillment of desire as the one who promises to fulfill our needs is able. Look toward the finish line. These are things that I am looking toward.

As the line approaches and the tape is about to snap. I lunge ahead.

April

★ 1 Apr 09

I am still awaiting my flight to the countries that lead me to the USA. The flight has been delayed for a couple of reasons, so I choose to watch more movies and sleep. Waiting and waiting is all I am doing at this moment. Delay after delay and time shift after time shift is most saddening. I just sit still and pray the Lord will speed up the process.

The process has stayed the same and did not move. It must not have been in God's will. The Word says ask and it shall be given, and it shall be done if it is God's will. That is the thing about prayer that many fail to realize. God must will it.

The purpose of seeking God and drawing near to him is for us to learn of him and begin to know him. Then our prayers can line up with his purpose and, we will see his manifestations. Healing, miracles, and power come through our obedience to God and our ability to discern his voice. It takes much practice. Like cheat meals, when we begin down the wrong path, so much work can be destroyed a lot more quickly than the time it takes to build. A bulldozer can destroy a home in minutes that took days and months to build.

As I approach the time of coming home, I am ready to hit the ground running. Catch up on my relationship with my wife and find out where I will be working when I return. In the military, sometimes your job is no longer open upon returning from deployment. This can be good if you did not like where you worked;but if you loved the place where you worked, you will not desire to leave.

Here I still sit, awaiting my travel.

★ 3 Apr 09

I am at the BWI terminal awaiting my brother to pick me up. I will stay the night at his house, as I have layover until tomorrow morning. First stop, McDonald's—the place of golden fries and a nice quarter pounder with cheese.

My brother and his wife came to get me, and we headed to McDonald's where I enjoyed a pleasant meal. I am ready to go to my home in Texas, but it is nice to be back in the US.

My brother and I played football, and then he showed me his latest karate flick. Unfortunately, I didn't get to watch much because I shortly went to sleep while it was playing. In the morning, he made me some pancakes, and we spoke a little about the Word. We spoke about how I tell him to use spellcheck and he tells me I have missing in my writing. It was a blessed time.

We prayed, and I felt my brother's anointing. It was awesome. How good it is to dwell together in unity. I know that some of you may not have a good relationship with your siblings. Remember,the Bible says there is a friend that sticks closer than a brother. So always be encouraged in the midst of family issues.

I arrived at the terminal with just enough time to go through the wickets of security checks and move to my seat. The plane ride was smooth, and we arrived in Dallas right on time. Dallas is my last leg, and it feels good. Soon, I will be home.

My number was called for travel to San Antonio, and while I buckled up, the flight attendant summoned me to first class. So I grabbed my bags and moved up. It is so nice. This is the first time I have ever flown first class. It's a big deal for me; others, something small.

Now that I am flying with room for my feet, and sipping from a real glass, I relax for the final stretch.

I arrive at the airport greeted by my wife. The timing of my arrival was unclear for my co-workers, so they did not show; but for this, I am glad. I like the simplicity of coming home and enjoying things familiar and all things that are family.

★ 4 Apr 09

As I close this journey through the desert, I reflect on answered and unanswered prayer. When I say unanswered, I simply mean not answered the way I desired it to be answered. In fulfilling my duty as a soldier, I will press on with whatever command is given, but my personal desire was for a shorter deployment. This was not answered amidst my complaints ushered to the Lord.

I asked God knowing he may not answer this prayer in the way I would have liked. I tried to ask it with as much faith as possible, but God has a will of his own and his will trumps all other wills. We must pray, expecting our Father who is in heaven to answer. We must also know that our Father chooses what to do based upon his unlimited perspective and not our limited vision. His thoughts are above our own. We must never go to prayer without a fervor to have our prayer answered.

Paul went before the Lord three times asking God to remove his thorn. Daniel prayed three times a day. Hannah prayed several times for a son. Abraham interceded several times for Lot's deliverance from Sodom. David fasted and prayed for his child to live. When the child died, which took a few days by the way, he got up, washed himself, and ate. God said no. David accepted it.

In the desert, I prayed for a shorter deployment and received one slightly longer. However, I can say that the perks I received as an E5 were greater than the perks received by those who out ranked me and I attribute this to God and his way of answering my prayer another way.

We must go to God without ceasing until the prayer has been answered. When you hear or see nothing, pray some more.

This Journey Ends

★ ★ ★

This journey ends in the desert with me learning valuable lessons—lessons that shall carry me to a possible new journey through another desert. I thank the Lord for those lessons, for prayers received by my wife, father, mother, brothers, sister, and church family. I have added to the arsenal of brothers in Christ, and it is good for brethren to dwell in unity together.

God's grace and mercy endured with me to the end, making difficult situations manageable and others easy as pie. The God we serve may test us and allow trials and tribulations, but be of good cheer. He has overcome the world and we can do all things through Christ which is our strength.

Call upon the Lord Jesus Christ today and call often. He did not fail me in the desert and he surely will not fail you.

Continue the journey with me at

RogerTharpe.wordpress.com

Contact me at

nimnez@aol.com

Made in the USA
Monee, IL
04 November 2026

67382378R00049